PIONEER VALLEY EDUCAT

WHAT IS A
FRACTION?

SEAN FINNIGAN

TABLE OF CONTENTS

Here is a pie.

How many pieces of pie do you see?

What if there is
just one piece of a pie?
How do we count that
one piece?

Fractions help us
count a part of a whole.

This cupcake is cut
into two equal parts.

When you cut
one whole cupcake
into two equal pieces,
you cut it in half.

This is how we write it.

$$\frac{1}{2}$$

We say we have
one half of the cupcake.

Here are some apples.
Can you find the apple
that is cut in half?

Here is one half of an apple.

This pizza can be cut
into three equal parts.

If we have just one piece
of the pizza, it is one piece
out of three pieces.

This is how to write it.

$$\frac{1}{3}$$

We say we have one third
of the pizza.

Here are some tomatoes. Can you find the tomato that is cut into thirds?

Here is one third of a tomato.

A peach can be cut into four equal parts.

If we have just one piece
of the peach, it is one piece
out of four pieces.

This is how to write it.

We say we have one fourth
of the peach.

Here are some limes.
Can you find the lime
that is cut into fourths?

Here is one fourth of a lime.

So what is a fraction?

A fraction is a part

of a whole

that is cut

into equal parts.

1
whole

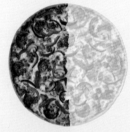

$\dfrac{1}{2}$

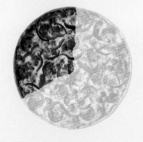

$\dfrac{1}{3}$

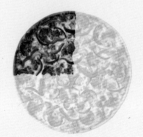

$\dfrac{1}{4}$